WHAT DO YOU MEAN BY 'AVERAGE'?

Means, Medians, and Modes

MEAN
MODE
MEDIAN
AVERAGE VOTERS UNITE!
Jill Slater for S.C. Pres!
TO REPRESEN
EVERYONE
We need the
AVERAGE!
Elect Jill Slater
THE ALL-AROUND AVERAGE CANDIDATE!

Elizabeth James & Carol Barkin

WHAT DO YOU MEAN BY 'AVERAGE'?

Means, Medians, and Modes

Illustrated by **Joel Schick**

Lothrop, Lee & Shepard Company

A Division of William Morrow & Co., Inc. New York

 Inquiries should be addressed to Lothrop, Lee & Shepard Company, 105 Madison Ave., New York, N. Y. 10016.
Printed in the United States of America.

First Edition 1 2 3 4 5 6 7 8 9 10

Library of Congress Cataloging in Publication Data

James, Elizabeth.
What do you mean by "average"?

SUMMARY: Defines and explains how to calculate mean, median, mode, and percentage.
1. Average—Juvenile literature. [1. Average. 2. Percentage]
I. Barkin, Carol, joint author. II. Schick, Joel. III. Title.
QA115.J35 513'.2 78-7227
ISBN 0-688-41854-6
ISBN 0-688-51854-0 lib. bdg.

By Elizabeth James & Carol Barkin

THE SIMPLE FACTS OF SIMPLE MACHINES

HOW TO KEEP A SECRET:
Writing and Talking in Code

WHAT DO YOU MEAN BY "AVERAGE"?

By Carol Barkin & Elizabeth James

SLAPDASH SEWING

SLAPDASH COOKING

SLAPDASH ALTERATIONS:
How to Recycle Your Wardrobe

SLAPDASH DECORATING

Contents

1 · What Do You Mean by "Average"?

We take you now to the sleepy little town of Normal City, where Jill Slater and her friends are having a meeting.

"I came up with a great idea for my campaign," Jill announces. "We're going to prove that I'm the most average person in our school, so I'll be the best person to represent everyone."

"That sounds terrific!" Harriet exclaims. "With that

campaign you're sure to be our next Student Council president."

"Yeah," David says, "but how do we prove it?"

"And what do you mean by 'average'?" asks Steve.

"Average" is a word you hear nearly every day. You may know the batting averages of your favorite ballplayers. You're sure to know that C is an average grade in school, but the average score on the last math test may have been 85. And, of course, everyone knows that the average student hates doing homework.

All of these are different kinds of averages. Averages are used to give information about all kinds of everyday things (not just in school!). But each kind of average gives different information. So when you hear that Brand X is best for the average headache or that the average laundry gets cleaner with Brand Y, you need to know which kind of average is being used. Averages can give you a lot of useful information, and they are convenient tools to describe a large number of individual facts about stubborn stains or anything else. But, like any tool, they can be confusing or even misleading if you don't know exactly how to use them. In our next installment, Jill's campaign committee finds out that there's more to averages than they suspect.

Tune in again. . . .

2 · The Mean

As we return to the sleepy little town of Normal City, Jill's campaign is waking a few people up.

"Hey, Jill, we got the lists of heights and weights for everyone in school!" David yells.

"Now what do we do?" asks Harriet.

"Well, get some pencils and a lot of scratch paper," Jill replies. "We have to add up all the heights and all the weights, and then divide the totals by the number of kids in school."

"Oh, no!" Steve groans. "I thought this was going to be more fun than that!"

Jill's campaign committee is finding the **mean** averages of height and weight for students in their school. A mean average can be computed for any collection of items that have numerical values: measurements, quantities, scores, and prices are often averaged. For example, when you plan a winter ski vacation, you may want to know the average snowfall for December in Utah or Vermont. On a cross-country trip, you may need to know the average

price of a hamburger at roadside restaurants as well as the average temperatures for the places you'll pass through (if it's August and the average temperature in Death Valley is 106° in the shade, plan a different route!).

Mean averages are always found by adding up the numerical values of all the items you're averaging and then dividing this total by the number of items. Because this kind of average is calculated from the numerical values, it is often called the **arithmetic mean.** To find the average daily high temperature for July, you first need to know the high temperature for each day in the month. These 31 numbers are added to get a total; the total is then divided by 31 to give the mean average high temperature for July. But for February, you would have only 28 temperatures to add, and you would divide the total by 28 (unless it's leap year!).

Back at campaign headquarters, Steve is knee-deep in scratch paper.

"Finally—that's finished!" he says. "Lucky for you, Jill, that I was always good at long division."

"Let's see the results," says Harriet. She reads off, "Average height—151 centimeters. Average weight—38 kilograms."

"Fantastic!" David exclaims. "That's exactly your height and weight, Jill!"

"I told you, I'm totally average." Jill smiles.

"Let's do another one," says David. "You're on the basketball team, Jill—are you an average player? What was your score in the last game?"

Players	**Scores**
Jill	8
Carmela	7
Diane	13
Jody	19
Arlene	10
Martha	0
Rosa	3
Cynthia	4
	$64 \div 8 = 8$

"Hey, look at that!" says Harriet. "If everyone on the team had made the same score you did, the total score would still have been 64."

A mean average represents a whole group of scores or other numbers: it is the score each person would have if the scores of the whole group were evened out. The individual scores themselves are called the **raw data.** The *range* of the raw data is the distance from the lowest numbers to the highest—in this case the range is 0 to 19.

The Mean

When you look at the raw data, you can see that Jill was the only player whose score was the same as the team's average. Knowing the mean average doesn't tell you how high or low any individual scores were, and it also doesn't tell you the range of the raw data. What it does tell you is the performance of the team as a whole.

"This averaging isn't too hard—we can use it for everything," says Harriet. "I've always wondered what the average allowance is—let's do that next."

Later that day:

"I have to tell you," says David, "the kids thought we were really nosy. But we found out how much allowance everyone in our homeroom gets."

"Good work, team," says Jill. "Let's tally up the results."

List of allowances:

$ 1.75
2.75
2.00
2.00
2.00
2.75
2.50
2.50
1.50
1.75
2.00
2.25 (Jill's)
2.75
3.00
2.00
2.50
2.00
2.75
1.75
25.00
2.50
2.50
1.50
2.00
2.75

total = $78.75

$78.75 ÷ 25 = $3.15 (mean average)

The Mean

"Oh, Jill, this is terrible," Harriet moans. "Your allowance isn't anywhere near the average!"

David looks up from his scratch paper. "This doesn't even make sense," he says. "There isn't anyone who actually gets $3.15 for allowance. Did we do something wrong?"

David doesn't have to worry—their calculations are correct. Since an arithmetic mean is found by adding and dividing a group of numbers, it often happens that the mean average is not the same as any one of the numbers used. You can easily see this for yourself:

$$\begin{array}{lll} 3 & 2 & 4 \\ 4 & 3 & 4 \\ \underline{5} & \underline{7} & \underline{4} \\ 12 \div 3 = 4 & 12 \div 3 = 4 & 12 \div 3 = 4 \end{array}$$

Remember that a mean average doesn't represent any individual item of the raw data; it is the number that each item would be if they were all evened out to be the same.

"Okay," says Steve, "but we have another really big problem. Only one person in the whole class gets more than a $3.00 allowance. I don't call that average."

"You've got something there," Harriet says. "If we

didn't have that rich guy Harvey in the class, the average would be much lower."

"Yeah, but we do have him," David says. "We can't just leave him out of the calculations."

The campaign committee has run into a real problem here. Sometimes raw data is spread out very unevenly: in the list of allowances, most are grouped in a small range from $1.50 to $3.00, while one item is much higher than all the others. This higher figure brings the mean average above the range of the other data. In any group of data, a few items that are very much higher or lower than the rest will pull the mean average up or down; for this situation, an arithmetic mean doesn't give the most useful information.

Most people think of an arithmetic mean when they hear the term "average," and a mean average is very useful for data that are fairly evenly spread throughout the range. But for raw data that are unevenly distributed, another method of averaging must be found.

As the sun sinks over the sleepy little town of Normal City, Jill snaps her fingers and shouts, "Aha! I've got the answer to our problem!"

Tune in again. . . .

3 · The Median

In the sleepy little town of Normal City, Jill Slater is still snapping her fingers. "We have to find a **median** average instead of a mean," she cries.

"What's that?" the campaign committee asks.

"It's the middle item in a group when the group is arranged in order," answers Jill. "Let's try it and see what we get."

(1) $1.50
(2) $1.50
(3) $1.75
(4) $1.75
(5) $1.75
(6) $2.00
(7) $2.00
(8) $2.00
(9) $2.00
(10) $2.00
(11) $2.00
(12) $2.00
(13) $2.25 median
(14) $2.50
(15) $2.50
(16) $2.50
(17) $2.50
(18) $ 2.50
(19) $ 2.75
(20) $ 2.75
(21) $ 2.75
(22) $ 2.75
(23) $ 2.75
(24) $ 3.00
(25) $25.00

The Median

"Hurrah!" cries Harriet. "Jill's allowance is average after all!"

"I don't get it," Steve complains. "How come it's the average this time but it wasn't when we found the mean?"

Finding a median average is really locating the center point in an ordered list of raw data. The order is usually based on numerical values—size, price, age, and so on—and it doesn't matter whether it is arranged from low to high or high to low. But unlike a mean average, a median can also be found for raw data that do not have numerical values. You can find the median color in a spectrum or the

median word or letter in an alphabetical list. As long as there is a way to arrange the data in order, a median can be located.

When you know the median average of a group, you know that in the range of the raw data there are as many items above the median as below it. This makes it very useful when your raw data are unevenly distributed, as in the list of allowances. The one allowance that is very different from the others cannot distort the picture of what's average for the group when you use a median. The U.S. Census uses a median to describe the average family income so that the few millionaires won't give a misleading impression of what the average family really has to spend. The mean average can give a distorted picture because it is based on the actual numerical values of the raw data. But a median is based only on position in the ordered group.

Don't assume, though, that the mean average the campaign committee found was wrong or completely useless. It told them that if all the allowance money for this class were split up evenly, each person would have $3.15 to spend. But that's not the information the committee was looking for. They wanted to know how much allowance the average class member really had to spend. And this information is what the median average provided.

As a researcher, it's important for you to know how

each kind of average works; then you can choose the one that will tell you what you really want to know.

David holds up a sheet of scratch paper. "Here's the list of family sizes for our class. What do we want to know about the average?"

"We want to find out how big a family the average class member really has," replies Harriet.

"That means a median!" Steve exclaims.

children per family	1 child	2 children	3 children
	////	~~////~~ /	~~////~~ ////
total number of children	4	12	27

children per family	4 children	5 children
	///	
total number of children	12	0

children per family	6 children	7 children
	/	//
total number of children	6	14

1 1 1 1 2 2 2 2 2 2 3 3 3 3 3 3 3 3 3 4 4 4 6 7 7

↑ median (the middle 3)

"Oh, good," Jill says. "I have two brothers, so I'm average again."

"Let's find the mean average too," suggests Harriet. "Then we'll also know the average number of children per family in this class."

"Good idea!" cries David. "Then we can even off the families and give away our extra brothers and sisters."

number of children

4
12
27
12
6
14

$75 \div 25 = 3$

"Look at that! The median and the mean are the same," says David.

Steve thinks a minute. "That must mean that the raw data are spread fairly evenly throughout the range."

As you know, the median locates the midpoint of a range of ordered raw data while the mean indicates the arithmetic average. If the distribution of the raw data is not too uneven, the mean and the median may be the same.

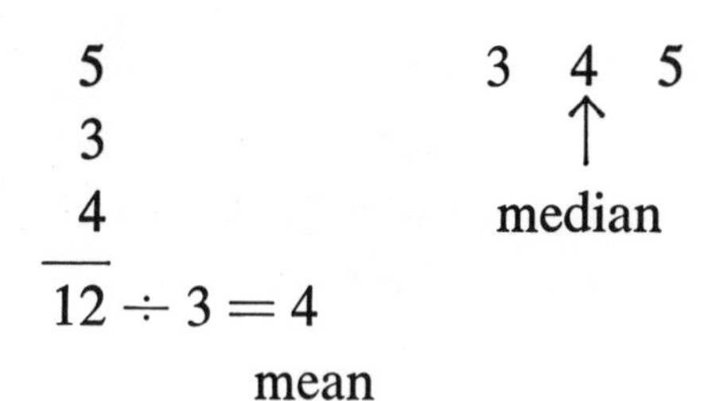

But you can see from the second example that when the data are spread very unevenly, these two kinds of average give you different information.

$$\begin{array}{r} 2 \\ 1 \\ 9 \\ \hline 12 \end{array} \div 3 = 4 \text{ (mean)}$$

1 2 9
↑
median

Remember that to find a mean, you need not put your raw data in order but you have to do some arithmetic; to find a median, no arithmetic is necessary but you do have to put the raw data in order.

Jill is pleased with the results so far. "I wonder whether it's average to be a girl in this class," she says. "Let's find out."

"Well, we can't use a mean," says David, "because

there aren't any numerical values to average."

Harriet says, "But there are fourteen girls and eleven boys. Why can't we add those figures and divide by two?"

"That won't give us any information we can use," replies David.

"Yeah, that's like averaging a red team with a blue team," Steve says. "You can only play on one team at a time. If you add two teams and divide by two, all you'll find out is how many players they'd have if both teams were equal. That's no help here."

"Okay, let's find the median. You don't need numerical values to do that kind of average," says Harriet.

Jill frowns. "We can't do that either. There's no way to put the data in order. After all, it's not like a color spectrum—how can we decide how to arrange them?"

"Let's figure out what question we're asking here," David says. "We want to know which is more common in this class—boys or girls."

Steve is puzzled. "That just sounds as if you're asking whether there are more girls or more boys. That's not an average, is it?"

"You've got it, Steve!" cries Jill. "That's exactly what we're asking here!"

Tune in again. . . .

4 · The Mode

Back in the sleepy little town of Normal City, Jill Slater's campaign committee is bewildered. Jill explains to them, "In this case we have to find the mode."

"What's that?" everyone asks.

"Whichever thing you have the most of in a group is the mode for that group," says Jill.

"You mean if you have three apples, two oranges, and

a banana, apples are the mode for that fruit bowl?" asks David.

"Yes," Jill replies. "So for our class, since we have fourteen girls and eleven boys, girls are the mode. I'm still totally average."

A **mode** or **modal average** is the item that occurs most frequently in a group. When you use the word "average" to mean "usual" or "most common," you are talking about a mode.

You have probably figured out that to find a mode, all you have to do is count. You don't have to put your raw data in order as you do for a median; so you can find a mode for data that have no natural order (like apples and oranges). You also don't have to do any arithmetic, as you do for a mean; so you can find a mode for data that have no numerical values.

Since the mode is defined as the most commonly occurring item in a range of data, it's obvious that the mode, unlike the mean, must always be the same as a real item in the raw data. But how about the median? Try finding the mean, median, and mode for this range of data:

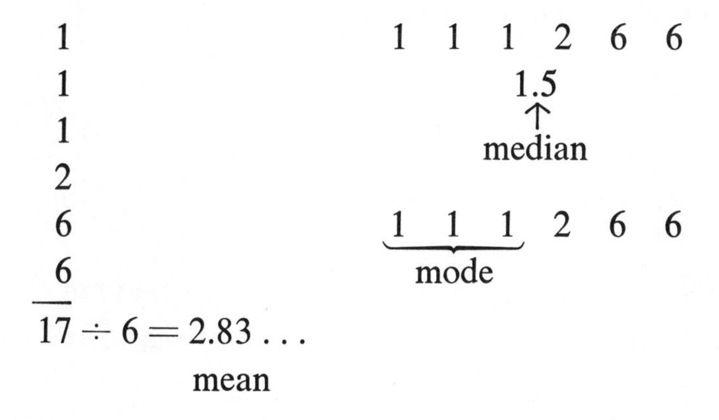

The mean average in this case is a number that does not appear in the data. You may remember that this also happened when Jill's campaign committee figured out the mean average allowance for their class. This is not an unusual result. But you may have been surprised that the median average of 1.5 also does not appear in the data.

You can easily see that this will happen whenever you have an even number of data. In this situation you must find the middle two items in the ordered data and then find the point that lies halfway between them. The mode, of course, is always an item that does appear in the data: here you have three ones, two sixes, and one two, so the modal average is one.

In the example above, the three averages are different from one another. However, look at the list of family sizes on page 23, and figure out the modal average. Here is a case where all three averages are the same.

"What other kinds of things can we average?" asks Steve.

"Too bad we can't do birthdays," David says. "Mine is tomorrow."

"Why can't we?" replies Harriet. "We can find out the modal month for birthdays. Let's look at the class list."

Jan	Feb	Mar	April	May	June
/	//		//	/	////

July	Aug	Sept	Oct	Nov	Dec
//	////	///	/	//	///

"Uh-oh," says Steve, "we're in trouble. There's no mode."

But Jill is smiling. "Oh, yes, there is. In fact, there are two modes—June and August. And my birthday is August fifteenth."

"Can you really do that? I thought there was only one average," asks Harriet.

Jill is right. There can be more than one modal average for a range of data. If two items occur the same number of times in a range, and no other items occur as often, then both these items are modes. Sometimes there are a lot of modes, but sometimes you may not be able to find any mode. How is this possible? When no items in the range of raw data are the same, there is no mode. If each person in your class gets a different score on a test, there is no modal score.

Modes can be used in a number of situations where a mean or median isn't very helpful. When a television commercial informs you that most people prefer Brand X toothpaste, do you recognize this toothpaste preference as a mode? Many products are advertised in this fashion. The advertisers are using "average" in the sense of "most common" or "most popular." So, the next time you hear that Yo-Yo Wax makes the average kitchen floor really sparkle, you'll know what to think!

"Well, Jill, it looks as if you really are average," says David. "Now what?"

"We have to find out what the average student thinks about my campaign issues. We'll do an opinion poll for the whole school," explains Jill.

Steve groans. "The whole school? We can't ask everybody—it'll take forever!"

"We'll never finish before the election—it's impossible!" Harriet moans. "Oh, Jill, what are we going to do?"

The curtain falls on a troubled campaign committee.

Tune in again. . . .

5 · Sampling

As the curtain rises on the sleepy little town of Normal City, Jill Slater looks around at her worried campaign committee. "Don't get upset," she says. "We don't have to talk to everyone in the whole school. We'll use a sample."

"Oh, great," says David. "That's what they do on television when they stop people in the street."

"How do we decide which kids to talk to?" asks Harriet.

A **sample** is a small part of a large group. To be useful, a sample must be chosen so that it gives a fair representation of the group as a whole.

Samples are generally used when the group being studied is very large. There are many reasons for using samples. It is expensive and time-consuming, for example, to question every bicycle rider or voter or toothpaste user in the whole country or even in one city. It's also difficult to train enough interviewers so that the questions will all be asked in the same way. (If one interviewer asks questions in a friendly way, he'll probably get different answers from the ones an angry, grumpy interviewer gets.)

If the sample is carefully chosen, it will provide an accurate picture of the whole group.

Before you choose a sample, you have to define the whole group you are studying. This is called the **universe.** (Of course, astronomers mean something different when they use this word!) An auto manufacturer, for example, may want to survey only people who own cars, not all the people who drive them. The information you want to find out from your survey will help determine the group you question (the universe). To find out whether the girls' locker room needs repainting, you wouldn't include any boys in the universe.

Once you have defined the universe, how can you make sure that the sample you choose really does represent this universe? This is the hardest part of designing any survey. You have to make sure that every person in the universe has the same chance of being included in the sample. If an auto manufacturer's sample includes only families with six or more children, it will probably show that everyone wants a large station wagon. This is a biased picture of the universe of all car buyers!

"Well, we know what our universe is—it's all the students in our school," Steve says. "Where's the list of questions?"

"Here's the questionnaire I made up," says Jill.

What should the Student Council raise money for next year?

1. a constellation globe for the library
2. new jerseys for the intramural soccer teams
3. new records to play at school parties
4. bicycle racks to accommodate more bikes

Should the Student Council try to:	**yes**	**no**	**don't care**
1. do something about noise in the lunchroom?			
2. provide music in the detention room?			
3. organize a yearly snowman contest?			
4. set up a swap center for exchanging lunch sandwiches?			

Harriet looks at the first question. "These are great ideas, Jill, but can the Student Council raise money for all those things?"

"No, we have to choose one of them," says Jill. "I think we need new records the most, but they're all good projects. It's up to the whole school to decide which one they want. That's why we need this survey question."

David is reading the second question. "Do you really want to have music in the detention room, Jill? I think that's weird."

"No," laughs Jill. "Only two of those questions are about *my* campaign platform—the lunchroom noise problem and the sandwich swap center. Those are the things I think need to be done. But Steve found out what my opponent Joe Blow is promising to work for—he wants a snowman contest and music in the detention room. So I want to find out whether the voters agree with him or with me."

"Okay. Let's use a sample of ten percent of the kids in school. There are four hundred kids, so we need to talk to forty of them," Harriet says.

"Sounds good," says David. "That should give us a fair idea of what the whole school thinks."

Steve nods. "We can each talk to ten people then. I'm going to the library after school tomorrow anyway, so I'll find ten people there. And David has soccer practice, so he can easily talk to ten of the players."

"Sure, and I'll get my ten at Drama Club tomorrow. Why don't you go to the detention room, Jill—there are always at least ten kids there!" exclaims Harriet.

Jill shakes her head. "We can't do it that way—that's not a fair sample. What about all the kids who do other stuff after school? They won't have any chance to be in the sample."

"That's true," says David. "And besides, everyone in detention is sure to want music. You can't have them be one quarter of the sample, since one quarter of the whole school probably doesn't go to detention."

Steve says, "Well, if one quarter of the sample represents one quarter of the school, maybe we should pick ten people from each grade. After all, each of the four grades has the same number of people so they all have an equal vote in the election."

"That sounds fair," says Jill, "but we still have to decide how to choose the sample from each grade."

David shrugs. "That's no problem. My brother is in seventh grade and his friends are always at our house. I'll ask ten of them."

"But, David," protests Harriet, "your brother and his friends probably all think alike about lots of these things. That wouldn't be a fair representation of the whole seventh grade."

"What are we going to do, then?" asks David.

"I know," says Steve. "We'll do what they always do on television—we'll take a **random sample**."

Jill's campaign committee has run into several common

problems in choosing a survey. But they have avoided some biases that might have weighted the sample in one direction or another. And by dividing their whole universe into four subgroups, one for each grade, they made sure that each equal group of the universe would have equal representation in the sample. Now the committee knows that the sample will have the same proportions according to grade as the whole school has.

Steve's suggestion is a good one—a random sample of each grade would be the best way to get an accurate picture of the students. However, a true random sample is difficult to obtain.

"Randomness" is a mathematical term. It means that in a group of random numbers, the numbers do not follow any order or pattern, and every number has the same chance of appearing as every other number. There are whole books of tables of random numbers that have been produced by computers; these tables must pass strict mathematical tests that guarantee that they are really random and free from bias. Using tables of random numbers to select a sample ensures an unbiased sample, but it is both difficult and time-consuming.

You might think that using such tables is not necessary, since you can just pick random numbers out of the air. But this is not really true. Everyone chooses some numbers more often than others and tends to arrange numbers in patterns or sequences. A list of numbers picked out of the air will almost never pass the tests for randomness.

Even a sample that is randomly chosen may produce results that are biased in unexpected ways. If every person chosen for the sample is not interviewed, you have only taken a sample of the sample, and you have no way of knowing how this smaller sample was chosen. If you called everyone at four o'clock in the morning, you probably got to interview a very small sample and it consisted only of real night owls. In fact, however your sample is chosen, you must make sure that this sample is the one you get and that it has not been redefined along the way by your interviewing methods!

"If we can't do a really random sample," says Harriet, "how can we choose which kids to talk to?"

"We need some kind of system," Steve says.

Jill pulls out some papers. "Here are the alphabetical lists of every student in each grade. Why not pick every tenth person on each list?"

"Sounds good," says David. "The alphabetical order of people's last names can't have very much to do with how they feel about these questions."

"Okay, we'll each take one grade," Harriet agrees. "But we all have to ask the questions the same way. It won't be fair for David to sound excited about the new soccer jerseys just because he plays soccer."

David grins. "Okay, Harriet, but we all know you're tired of baloney sandwiches every day. Be careful not to bias *your* results in favor of the sandwich swap center!"

Steve says, "Let's all remember that we have to track down every student who's in our samples—it won't be any good if we skip the ones who are hard to find."

"I guess we'll find out tomorrow how average my opinions are," says Jill. "I hope the voters agree with me about what's important."

As we leave the sleepy little town of Normal City, the weary campaign committee is preparing for a big day tomorrow.

Tune in again. . . .

6 · Percentages

Welcome back to the sleepy little town of Normal City, where a torrential rain is causing the worst floods in twenty years. Nevertheless, Jill Slater's campaign committee is full of enthusiasm as they tally up their survey results.

"Well, it wasn't easy," says Harriet, "but I finally found the last kid in my sample—he was floating a message in a bottle out the detention room window!"

"Okay, let's see what we've got," says David.

What should the Student Council raise money for next year?

1.	a constellation globe for the library	6
2.	new jerseys for the intramural soccer teams	8
3.	new records to play at school parties	22
4.	bicycle racks to accommodate more bikes	4

Should the Student Council try to:	**yes**	**no**	**don't care**
1. do something about noise in the lunchroom?	19	12	9

2. provide music in the detention room?	6	10	24
3. organize a yearly snowman contest?	18	14	8
4. set up a swap center for exchanging lunch sandwiches?	29	5	6

Harriet looks at the tally sheet for the first question. "Jill, you're sure right on this one. Most of our sample thinks that new records are what we really need."

"Then that's the mode," says Steve. "That means the average voter agrees with Jill."

"I can see that more than half of the sample wants new records, but how big a majority is it?" asks David.

As David's question indicates, a mode tells you which item is most common or popular, but it doesn't tell you how *much* more popular it is than the other items. You have no way to compare the popularity of all the items.

An easy way to look at all your results and compare them with one another is to convert your raw data to percentages. "Per cent" means "per hundred," or "hundredths." Of course, your total number of responses is 100 percent, so the number of responses for each choice is a portion of 100 percent.

When you ask what percentage of the sample wanted new records, you are really asking how many hundredths are equal to twenty-two fortieths, or

22 is to 40 as ? is to 100.

You can write this problem as:

$$\frac{22}{40} = \frac{X}{100}$$

To find out what X equals, multiply both sides of the equation by $\frac{100}{1}$:

$$\frac{22}{40} \times \frac{100}{1} = \frac{X}{100} \times \frac{100}{1}$$

$$\frac{2200}{40} = X$$

$$2200 \div 40 = 55\%$$

An easier way to find the percentage is to divide the smaller part by the total and convert the decimal to percent:

$$40\overline{)22.00}^{\;.55} = \frac{55}{100} = 55\%$$

Try finding what percent of the sample preferred the other three choices, and see if your answers are the same as the campaign committee's. (Hint: a good way to check your division is to see if your percentages add up to 100 percent.)

"Here are the percentages for the first question," Jill says.

1. constellation globe	6	15%
2. soccer jerseys	8	20%
3. records	22	55%
4. bike racks	4	10%
	40	100%

"This is great, Jill!" cries Harriet. "Fifty-five percent of the voters agree with you, and you only need fifty-one percent to win the election!"

"We're in good shape!" David exclaims. "Let's see how the second question comes out."

	yes	**no**	**don't care**
1. lunchroom noise	19	12	9
2. detention room music	6	10	24
3. snowman contest	18	14	8
4. sandwich swap center	29	5	6

Steve looks puzzled. "There are so many numbers for this one. How do we find the percentages?"

"I guess we really asked four separate questions here," replies Jill, "and each question has three possible answers. We have to find out what percentage answered 'yes,' 'no,' and 'don't care' for each question."

David groans. "Get out the scratch paper!"

1. lunchroom noise

yes	19	47.5%
no	12	30.0%
don't care	9	22.5%
	40	100%

2. detention room music

yes	6	15%
no	10	25%
don't care	24	60%
	40	100%

3. snowman contest

yes	18	45%
no	14	35%
don't care	8	20%
	40	100%

4. sandwich swap center

yes	29	72.5%
no	5	12.5%
don't care	6	15.0%
	40	100%

Percentages

"Wow, Jill, your sandwich swap center idea is a real winner! Seventy-two and a half percent!" shouts Steve.

"Yes, and a majority agrees with you on doing something about lunchroom noise," says Harriet.

"But it's not a majority of over fifty percent," replies Jill. "Who knows how the 'don't cares' will vote?"

"And look at the snowman contest percentages," David says. "Joe Blow has almost the same percent who like his idea there."

"It's awfully hard to look at these numbers all at the same time," says Harriet. "Isn't there some way to make it easier?"

Harriet has discovered how confusing it can be to compare a bunch of numbers. To avoid this confusion, you can use graphs to give everyone a clear picture of your findings. Jill's first survey question results can easily be shown in a circle or pie graph:

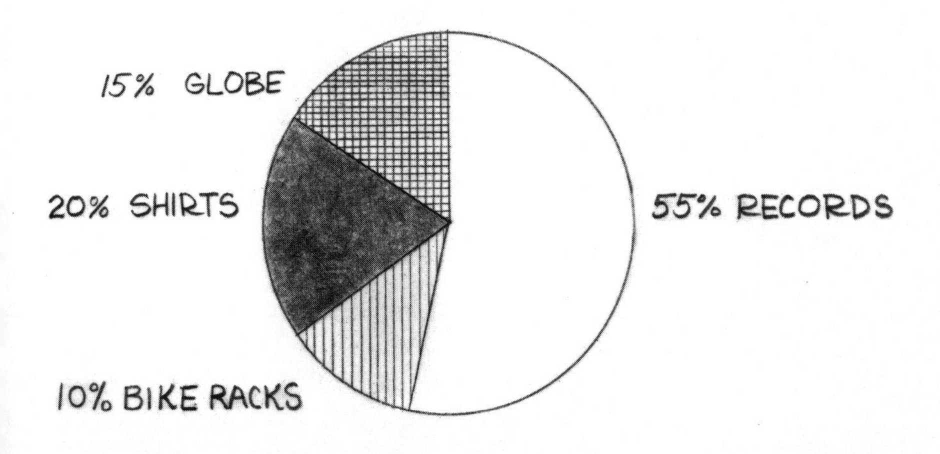

This is a good method for a question that has many possible answers.

Four different pie graphs could be used for the four separate parts of Jill's second survey question, but it would still be hard to make comparisons. Another common type of graph—a bar graph—would make the comparisons much clearer.

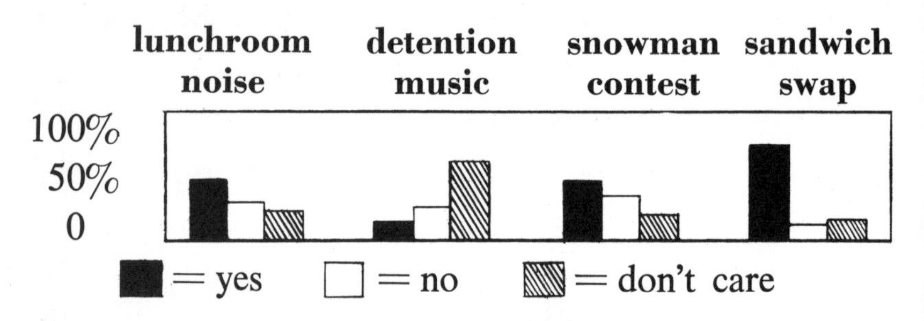

When you are comparing several similar questions, a bar graph is easier to read than a series of pies.

Graphs showing percentages are used frequently in newspapers and books to illustrate many kind of statistics. Using percentages makes it possible to translate various kinds of raw rata into a universal language.

Another common use of percentages is in percentiles. You may have been told that you are in the 87th percentile for height. This means that 87 percent of people your age are shorter than you are. It does not mean, however, that your height is 87 percent of what it will be when you stop growing, or that you are 87 percent taller than anyone else.

Percentiles can only be used for raw data that can be put in order from high to low—height, weight, test scores, and so on. Instead of comparing your performance on a test with the best possible or perfect score, percentiles compare your performance with that of everyone else who took the test. So, if you are in the 87th percentile on a national achievement test, it doesn't mean that you got 87 percent of the answers right. What it tells you is that you did better than 87 percent of the people who took that test.

Percentiles provide a way of comparing the members of a large group to one another. But they can be used only for data that can be put in order. Jill's campaign committee could not have found percentiles for their sample, because opinions and yes-no answers have no order.

"Well, Jill, these percentages look pretty good," says Steve.

"But you don't have over fifty percent that agree with your ideas on everything," David says. "There are a lot of 'don't care' votes that we can't predict."

Harriet nods. "You'd better get to work on your campaign speech, Jill. You have to persuade a lot of those 'don't cares' to agree with you at the election assembly."

"I know," says Jill. "I think I could do a good job as Student Council president, but unless my speech really says what I feel about these issues, I may not have a chance to find out."

Will Jill find a way to capture the votes she needs?

Tune in again. . . .

7 • What It All Means

Our last visit to the sleepy little town of Normal City finds Jill Slater's campaign committee waiting for Jill.

"Boy, our posters really look fabulous!" Harriet exclaims. "David, that sandwich board you had on at lunch was incredible!"

David smiles. "It was really fun. Everybody came up and took a look, so they all found out about Jill's sandwich swap center idea."

"I like the campaign slogans we came up with," says Steve.

AVERAGE
VOTERS
UNITE!

Jill Slater
for
S. C. Pres.

TO REPRESENT EVERYONE
WE NEED THE AVERAGE

Elect Jill Slater, the all-
around average candidate

"Me too," Harriet says, "and I love the posters you made with the graphs showing Jill's averages. The campaign is going really well."

"I wish Jill would hurry up and get here," David says. "I want to hear her rehearse her speech. It should be great."

Crash! The door flies open and Jill rushes in. "You won't believe what I just found out! The school is planning to pave over the garden behind the library and make it into a parking lot! I'll have to scrap my whole speech and start over—this is a really big issue!"

"Oh, no!" Steve groans.

"You can't do that, Jill," says Harriet. "We don't have time to do another survey to find out how the students feel about it."

"All that hard work for nothing," moans David.

"It wasn't for nothing," replies Jill. "We found out that the average person agrees with me on other issues, so maybe they'll agree about the garden too. In my speech I'll try to persuade everyone that it's important to keep our garden. But I can't just ignore this issue—I have to take a stand for what I think is important."

As Jill rushes out to work on a new speech, her campaign committee workers exchange worried looks.

In spite of careful planning, Jill and her committee have run into an unexpected problem. This sometimes happens even in complicated scientific experiments. New information may become available after the research is complete. Or a researcher may think that he has planned for every possibility only to find after months or years of work that he's overlooked some important factor. His results in this case will not mean what he thought they would. Some people are tempted to ignore this problem and pretend that their results are accurate. But, of course, any conclusions are only as good as the methods used to find them.

When you look at someone else's research results, you don't know what methods were used to find them. When an ad claims that four out of five people surveyed prefer new Super Minty Toothpaste, it may not mention that the sample included only big peppermint eaters. And have you ever asked yourself, "They prefer Super Minty to what—plain baking soda, their regular mint toothpaste, or a new toothpaste that tastes like garlic?" Without knowing who was asked what question, it is impossible to know what the results are worth.

Of course, Jill and her committee didn't have this problem. For each kind of question they were careful to use a method that gave them meaningful results. They not only learned to find means, medians, modes, and percentages, but they also found out how to choose the best method for each situation. The only problem they have is that the new information came too late, so their research didn't cover everything they now want to know.

Jill and her campaign committee are sitting in the garden outside the school library awaiting the election results.

"Your speech was wonderful, Jill," says Harriet.

"Yeah," David adds, "you really convinced me about saving this garden."

Jill crosses her fingers. "I just hope I convinced the average voter."

"We'll know before long. They're counting the ballots now," Steve says.

Just then the library door opens and Bob, this year's Student Council president, comes out.

"Gosh, good luck, Jill," whispers Steve.

Bob smiles. "Hope you'll like the job, Jill. You'll have it for a whole year."

Jill jumps up. "You mean I won?!"

"By a landslide—ninety-seven percent!" says Bob.

"Congratulations, Jill!" shouts David. "But now that you're president, you're not average anymore."

"Oh, yes, I am," says Jill. "It just goes to show what the average person can do!"

Index

About the Authors

Elizabeth James (Mrs. David Marks) received her B.A. in mathematics from Colorado College, where she minored in experimental psychology, and continued her studies at UCLA and California State College at Long Beach. A consultant in educational television, she has written scripts for both theatrical and documentary films as well as educational television programs. She and her husband live in Beverly Hills, California.

Carol Barkin received her B.A. from Radcliffe College, where she majored in English. She and her husband have traveled widely in Europe, the Far East, and the United States, and became friends with Elizabeth James when they lived for a time on the West Coast. In addition to editing and writing for a children's periodical, Ms. Barkin has had several years' experience editing adult and children's books in the United States and in London. She and her husband now live in New York City, with their young son.

About the Illustrator

Joel Schick has illustrated many children's books, including *How to Keep a Secret: Writing and Talking in Code* by Elizabeth James and Carol Barkin and several by his wife, writer Alice Schick. He and his wife live in Monterey, Massachusetts, with their young son, a dog, six cats, and two gerbils.